THE POETRY OF LEAD

The Poetry of Lead

Walter the Educator

Silent King Books

SILENT KING BOOKS

SKB

Copyright © 2024 by Walter the Educator

All rights reserved. No part of this book may be reproduced in any manner whatsoever without written permission except in the case of brief quotations embodied in critical articles and reviews.

First Printing, 2024

Disclaimer
This book is a literary work; poems are not about specific persons, locations, situations, and/or circumstances unless mentioned in a historical context. This book is for entertainment and informational purposes only. The author and publisher offer this information without warranties expressed or implied. No matter the grounds, neither the author nor the publisher will be accountable for any losses, injuries, or other damages caused by the reader's use of this book. The use of this book acknowledges an understanding and acceptance of this disclaimer.

> "Earning a degree in chemistry changed my life!"
> – Walter the Educator

dedicated to all the chemistry lovers, like myself, across the world

LEAD

In shadowed depths where silence reigns,

LEAD

Lead's heavy heart, in darkness chains.

LEAD

A metal forged in earth's embrace,

LEAD

It bears the weight of time and space.

LEAD

LEAD

In veins of rock, it lies concealed,

LEAD

A silent sentinel, its fate sealed.

LEAD

With stoic grace, it stands alone,

LEAD

In realms where sunlight ne'er is shown.

LEAD

From ancient mines, its essence drawn,

LEAD

A silent witness to the dawn.

LEAD

Through alchemy's arcane embrace,

LEAD

Lead takes its form, its destined place.

LEAD

In molten streams, it flows and melds,

LEAD

A liquid dance, where truth compels.

LEAD

Its shimmering surface, dark and cold,

LEAD

A mirror to the tales untold.

LEAD

In bullets cast, in soldiers' hands,

LEAD

Lead marks the course of war's demands.

LEAD

A metal of both life and death,

LEAD

It steals the very breath of breath.

LEAD

Yet in its silence, there lies a power,

LEAD

To shield, to shape, to build a tower.

LEAD

For leaden sheets, in roofs they lie,

LEAD

Guarding against the stormy sky.

LEAD

In pipes it flows, in lines it's drawn,

LEAD

Delivering water dusk to dawn.

LEAD

A humble servant, hidden true,

LEAD

In service to the many, not the few.

LEAD

In pencils, too, it finds its place,

LEAD

A tool of art, a vessel of grace.

LEAD

With graphite's mark, it leaves its trace,

LEAD

On canvas white, in endless space.

LEAD

Lead's journey long, its story vast,

LEAD

From ancient mines to present mast.

LEAD

A testament to time's embrace,

LEAD

A symbol of both strength and grace.

LEAD

So let us honor, let us heed,

LEAD

The silent call of lead's creed.

LEAD

For in its depths, there lies a tale,

LEAD

Of resilience that will never fail.

LEAD

ABOUT THE CREATOR

Walter the Educator is one of the pseudonyms for Walter Anderson. Formally educated in Chemistry, Business, and Education, he is an educator, an author, a diverse entrepreneur, and he is the son of a disabled war veteran. "Walter the Educator" shares his time between educating and creating. He holds interests and owns several creative projects that entertain, enlighten, enhance, and educate, hoping to inspire and motivate you.

Follow, find new works, and stay up to date
with Walter the Educator™
at WaltertheEducator.com

www.ingramcontent.com/pod-product-compliance
Lightning Source LLC
LaVergne TN
LVHW051921060526
838201LV00060B/4112